CAREER AS AN

ELECTRICIAN

ELECTRICAL CONTRACTOR

MOST PEOPLE TAKE ELECTRICITY FOR granted – at least until a storm knocks the power out. Playing Monopoly by candlelight instead of Minecraft on your Xbox is entertaining for a while, but it gets old fast when you are accustomed to flashy video games. If you stop and think about it, we are all very dependent on electricity. We all think of air conditioning, lights, and electronic gadgets as basic necessities. Without them, life would be cold, dark, and boring.

Electricians are the people who keep us plugged into the conveniences of modern life. Most of us have heard of circuits, fuses, volts, and watts. Electricians actually know what they are and what to do with them. There are around 600,000 professionals in this field who know the ins and outs of designing, installing, and repairing electrical systems. They read blueprints, solve math problems, and make sure the electrical work is up to code.

Most electricians work in one of two areas: construction or maintenance. Those in construction wire entire buildings, whether they are residential homes, office buildings, or factories. Their work is usually done during new construction, though some may work on remodels or upgrades for old existing systems.

Maintenance electricians always work on existing systems or equipment. Most are employed by large businesses or factories that need someone to keep machines up and running, often around the clock. Maintenance electricians spend their time testing and inspecting electrical systems, controls, and motors to avoid outages and costly downtime.

A growing number of electricians are choosing to focus on certain specialties. The specialty may be an industry, such as iron and steel mills or automotive plants. It could be a particular type of work such as servicing residential customers or converting businesses to alternative energy sources.

This is one of the few careers that requires only a high school diploma to enter a high paying field with a solid future. Most electricians start out as apprentices, getting paid on the job as they learn. It takes about four years to complete the training and qualify for whatever licensing the state requires During that time, automatic pay raises come along as the trainee gains knowledge and experience. Most apprenticeships are offered by a union such as the National Electrical Contractors Association, but nonunion companies offer similar training programs to beginners.

The job outlook for electricians is excellent. Analysts predict there will be around 225,000 new positions opening up over the coming decade. America is becoming more dependent than ever on consistent and reliable supplies of electricity. The area of alternative and renewable energy is expected to boom. Electricians who take additional classes to learn more about this fascinating new field will be in the best position to take advantage of new opportunities in green energy.

If you enjoy working with your hands while exercising your brain, this career is worth a look. It offers exceptional job security, good pay and benefits, and a wide array of specialties that can satisfy just about any special interest you may want to pursue.

WHAT YOU CAN DO NOW

THERE IS MUCH YOU CAN DO TO prepare for a career as an electrician while still in high school. In fact, you can even start your career before graduation, but do not ignore the opportunities to get an education that will help you advance in this career. Some of the courses that you should include in your curriculum are algebra, geometry, physics, chemistry, workshop (any kind), English, and mechanical drawing. If you are the independent type who might want to set up your own business someday, also include any business and computer courses that might be available.

Some cities have vocational high schools. These schools provide the same academic foundation as ordinary high schools, but the focus is on preparing students for a particular vocation. You could graduate with a high school diploma and an electrician certificate, and step directly into a career that will earn you more than many college graduates can expect.

If there is no vocational school near you, it is still possible to earn an electrician certificate by attending an online school. The certificates obtained this way are just as valid as those from regular schools – if the school is accredited by the relevant accreditation boards. Be sure to check any school's accreditation status before you apply.

Joining a high school electronics club will provide you with additional background. Volunteering for organizations such as Habitat for Humanity can also provide you with invaluable experience.

Not sure you want to be an electrician? Job shadow an electrician on the job for a day. This is a great way to explore whether it is the right career for you. Ask your school counselor to help you set it up. Be sure to ask plenty of questions and gather advice on things you can do to prepare.

If you are at least 16 years old, you can work at a part-time job after school or during the summer. Some states will require you to obtain an Electrician Trainee certificate, but it is a simple application and the cost is nominal.

HISTORY OF THE CAREER

MOST PEOPLE THINK ELECTRICITY WAS invented by Benjamin Franklin. It is a popular myth, particularly in the United States. The fact is that the history of electricity dates back much further. Greek philosopher and mathematician, Thales of Miletus, wrote about static electricity in 600 BC. He referred to the phenomenon as "charging amber." In 1660, German scientist, Otto Von Guericke, built a machine that was able to create static electricity. Neither had any idea that this force of nature could be harnessed and distributed to precise destinations through conductivity. That discovery was not made until 1720. A short time (10 years) later, a French chemist, Francois Due Fay, discovered that there are two different forms of electricity: negative and positive. He gave them names that are more elegant: "vitreous" and "resinous."

This brings us to Benjamin Franklin, who came up with a new theory about electricity in 1752. His theory was a bit risky to test as it required him to expose himself to the danger of a lightning storm. Every American schoolchild knows the story of Benjamin Franklin's experiment, how he flew a kite with a metal key attached to the string. The outcome of this experiment proved Franklin's theory that lightning is electricity. For this, he was awarded a Copley Medal, the Royal Society's oldest and most prestigious award given only to the most notable scientists. Identifying electricity was a great step forward, but it would be another 100+ years before the world figured out what to do with it.

In the 1870s, Thomas Edison built the first direct current (DC) electric generator. The first public electricity supply was not created until 1881, and it did not even use Edison's "technology." Instead, it was rather crude, involving a waterwheel from a mill. Edison's system was eventually used to power all of New York City. Although the problem of distributing electricity had been solved, there was still the matter of what it would be used for. Within a few years, the most important inventions involving electricity came along. Alexander Graham Bell invented the telephone, which used electricity to transmit speech across great distances.

Thomas Edison created a successful carbon incandescent lamp. This invention is credited with changing the course of history. Electric lighting, introduced in the late 19th century, was the first widely used application of electrical energy. Many other developments in electrical usage for homes and factories soon followed. There was a rapid expansion in the use of electrical technology, as scientists and others came to understand the immense versatility electricity offered as an energy source.

The electrician trade was born out of the need for experts who could install, repair, and maintain electrical systems and appliances. Many of the first electrical contractors rose from the ranks of employees in the Edison Electric Illuminating Companies that were formed around the country to provide electricity for homes and businesses. Other early electricians began by working for the telephone companies or for the forerunners of public utilities, which were set up primarily to power electric streetcars. Most of the action was found in big cities such as New York, and that is where the first independent electrical contracting business opened, in 1882. Soon after, there were hundreds of other electrical contracting companies started in major cities all over the country.

Trade associations began to form around the turn of the century. Local associations of electrical contractors offered a place for individual contractors to meet and exchange ideas. In 1901, a group of electrical contractors met at the Pan-American Exposition in Buffalo, where a major display of electric power had been installed. The electricians agreed to join together to nurture trade opportunities and establish guidelines for the fledgling industry on a national basis. The organization continues today as the National Electrical Contractors Association.

At the turn of the 20th century, electrical inventions and innovations, from the esoteric to the mundane, were introduced at a breathtaking pace. Albert Einstein invented photovoltaic cells, which demonstrated the ability to create energy using solar power. In 1918, the first washing machines and refrigerators came rolling off the assembly lines. By the 1930s, electrical appliances had become a part of every household.

By the 1950s, electrical usage was ubiquitous in the US. Research and technology led to large scale electrical systems like networked power grids and nuclear power stations, and extended to electronic engineering of minute electronic circuits.

Specific expertise was needed in many different areas, a

fact that led to electricians specializing in certain types of installation, maintenance, and repair. By the end of the 20th century, the electrician trade was essentially divided into three different types: domestic, commercial, and industrial. Each type required electricians to learn the particular skills needed to work within that area.

The future has never looked better for electricians, especially those who are ready to break into new and exciting fields. Scientists and engineers are finding new ways of using electricity as an energy source, from electric cars to remote home monitoring systems. More importantly, new sources of electricity are moving beyond the experimental stage and into the mainstream. Wind and solar farms are plugging into the national power grid.

The world's first commercial wave power station has been built on an island off Scotland. This unique invention generates electricity by using wave motion on the shoreline to compress air, which drives turbines to create power. This single wave power station can provide electricity to over 400 homes – and it is absolutely clean and renewable.

WHERE YOU WILL WORK

THERE ARE ABOUT 600,000 ELECTRICIANS working in the United States today. You can find them anywhere electrical power is being used – which is just about everywhere. Some are employed by various government agencies to maintain power systems, build huge power plants, or manage power-producing dams. Even the smallest municipalities need electricians to keep street lamps and traffic lights working.

The largest numbers of electricians, roughly two-thirds, are employed in the construction industry. They can be found working indoors and outdoors, in homes, businesses, factories, and construction sites of all kinds. Most electricians work close to home, but those in the construction industry sometimes have to travel to different work sites. Local or long distance commuting is not uncommon.

Maintenance electricians work primarily in factories, making sure the power systems keep the equipment and assembly lines functioning. This is some of the highest paid work an electrician can get, but it can mean being subjected to noisy machinery.

About 10 percent are self-employed. They take on projects big and small from general contractors and individuals who need help on a temporary basis. The work for these electricians can vary from day to day or even hour to hour. The work can involve just about anything associated with electricity from rewiring an old house, to bringing it up to code, to installing a home theater.

Many electricians work alone, but at larger companies and on big projects, they are more likely to work as part of a crew. Journeyman electricians may direct helpers, or train and guide apprentices.

Almost all electricians work full time. Work schedules may include evenings and weekends if shiftwork or construction deadlines are involved. Work schedules may also vary during times of inclement weather. Self-employed electricians often have the ability to set their own schedule.

THE WORK YOU WILL DO

ELECTRICITY IS ESSENTIAL FOR LIGHT, power, security systems, temperature control, and communications. Electricians install, connect, test, and maintain electrical systems for a variety of purposes in homes, businesses, and factories. Almost every building has an electrical power system that is installed during construction and maintained after that. These systems power the lights, appliances, and equipment that make people's lives and jobs easier and more comfortable.

Electricians were originally people who demonstrated or studied the principles of electricity. Today, they are skilled professionals who are trained in the electrical wiring of buildings, stationary machines, and related equipment. Not all electricians work in buildings though. Some specialize in wiring ships, airplanes, and other mobile platforms.

There are numerous guidelines that electricians need to follow during installation. These include local and state building codes and the National Electric Code. Learning and staying current with changes to these rules are a big part of every electrician's training.

Electricians work with specifications or blueprints when they install electrical systems in factories, office buildings, homes, and other structures. Blueprints are like maps that show where circuits, outlets, load centers, panel boards, and other equipment are located or need to be located.

When electricians start installing electrical systems in factories and offices, they first place conduits inside partitions, walls, or other concealed areas. Conduits are pipes or tubing that will house wiring. In lighter construction, which is typically residential housing, plastic-covered wire is used instead of heavy conduit.

Small metal or plastic boxes are then attached to the wall. These boxes will house the necessary switches and outlets. Insulated wire or cable is pulled through the conduit to complete the circuit between the boxes. The wire connects to circuit breakers, transformers, and other components. Simple hand tools are used for most of the work, such as screwdrivers, pliers, knives, and hacksaws. Power tools are also used occasionally.

Once the wiring is completed, the electrician uses test equipment, such as ohmmeters, voltmeters, oscilloscopes, ammeters, and test lamps to check the circuits for proper connections. This final inspection of the work is extremely important. The electrician is looking for possible flaws such as incompatibility with other systems and safety issues.

In addition to wiring a building's electrical system, electricians may install coaxial or fiber optic cable for computers and other telecommunications equipment. A growing number of electricians install telephone systems, computer wiring and equipment, street lights, intercom systems, and fire alarm and security systems. They also may connect motors to electrical power, and install electronic controls for industrial equipment.

The majority of electricians work in either construction or maintenance. A growing number do both.

Construction Electricians

Construction electricians work on both residential and commercial buildings. Most like to focus on larger projects, such as installing a new electrical system for an entire building, or upgrading an entire floor of an office building during remodeling. Working in new construction is favored because it is considered easier than retrofits or maintenance. That is because installing electrical systems in newly constructed buildings is less complicated than existing structures, and the wire is more easily accessible during construction. Plus, there is no need for troubleshooting in a system that has not been used yet.

Aside from a structure's entire electrical system, electricians might also install low voltage systems for video, information, and audio systems like telephones, internet connections, intercoms, and various types of alarms. They might also put in fiber optic cable or coaxial cable for computers and machinery controls.

Electricians often work alone, but in the construction field collaboration is necessary. For example, a master electrician may be asked to work closely with building engineers and architects to help design electrical systems before the new construction begins. Some electricians may consult with other construction specialists, such as elevator installers or HVAC (heating and air conditioning) contractors. At larger companies, electricians are likely to work as part of a crew.

Electrical Contractors

Most construction electricians are employed by contractors during the secondary phases of building. Electrical contractors are businesses that employ full-time electricians or contract freelancers on a project basis. The electrical contractor is responsible for generating bids for new jobs, hiring the electricians that will be needed to complete the work, and providing material and supplies to electricians as needed. They also communicate with architects, electrical and building engineers, and clients, from the planning stages through project completion.

Maintenance Electricians

Maintenance electricians are involved in the repair and upkeep of existing electrical systems. This work can be very different based on whether the electrician works in the residential or commercial field. Those who specialize in residential work are often known as service electricians. These workers might rewire an older house's electrical system or replace an old fuse box with a new circuit breaker when new appliances are installed.

Commercial or industrial maintenance electricians work as freelancers or are employed directly by large factories, office buildings, or hospitals. Those who work in large factories repair motors, transformers, generators, and electronic controllers on machine tools and industrial robots. Those in office buildings and smaller industrial facilities repair all types of electrical equipment from lighting to conveyors.

Industrial maintenance electricians focus their time on preventing problems from happening. They conduct routine tests and inspections of systems and equipment, make assessments, identify potential problems, and take steps to correct them. Before taking action that could lead to disruptions and costly repairs, they advise management on whether continued operation of equipment could be hazardous if servicing is delayed until a convenient time.

When breakdowns do occur, they must make the necessary repairs as quickly as possible in order to minimize downtime. Repairs may require replacement of circuit breakers, fuses, switches, electrical and electronic components, and wire. Sometimes it becomes necessary to install complete new electrical equipment.

Specialties

General electricians are at a disadvantage in this increasingly complex field. Wages and salaries are much higher than the average for those who get additional training and specialize in a particular area or pursue a specific type of electrical work. Some examples are:

Cable splicer

Data communications professional

Certified fiber optics specialist

Certified nuclear specialist

Tower climber

Electronic home entertainment equipment installer or repairer

Electrical drafter

Construction and building inspector

Lineman

Elevator installation and maintenance professional

Stage and theatrical lighting

Other popular specialty areas are marine electricians, research electricians, and hospital electricians.

Opportun ties in solar, wind, and other alternative energy are increasingly popular for those who want to specialize in green energy.

Computers and communications

Electricians are finding that their work often overlaps with computer and telecommunications wiring. Electrical and electronics systems are usually installed at the same time, and new structures are wired for networks and telecommunications immediately. For this reason, many electricians take additional classes on telecommunications systems, wiring, and the electrical interfaces so they can do this work themselves.

Consultants

Some experienced electricians work as consultants. They may make recommendations about the type of system a company might want to install or whether an existing system should be updated to increase safety or efficiency. Sometimes these consultants act as troubleshooters. When problems occur at a client's facility, they are called in to assess the situation and determine how to get the system up and running again as quickly as possible.

ELECTRICIANS TELL YOU ABOUT THEIR CAREERS

I Work in a Power Station

"The array of work an electrician can choose to do is enormous. It can be anything from simple house wiring to very complicated fiber optic connections. I have always preferred working on projects that have complex systems. I have worked at a wastewater treatment facility and on dams, and I was even a consultant to a wind farm.

My current employer is a large electrical contractor that specializes in emergency power systems for hospitals. My crew is in the process of installing several new unit substations. This is a perfect fit for me. I'm something of a perfectionist, and medical facilities require a great deal of accuracy. As crew leader, it is my responsibility to ensure that everything down to the smallest detail is done right.

This job is physically demanding, but my mind has to be sharp as well. It has taken a lot of training to learn how to do the work correctly. The kind of complex systems I work on could kill you in an instant. Some people assume that because they were able to install track lighting in their home, becoming an electrician is easy. It is not. There is a reason it took me four years of study to reach journeyman level. Anyone with average intelligence can be taught about volts and amperes during an apprenticeship. The kind of complex systems I work on involve a million details.

The best electrician is someone who is smart and can

think outside the box. Many of those who excel in this field are critical thinkers who can quickly diagnose problems. Patience is just as important as speed. It takes time to solve complex issues with equipment that is very expensive and demanding.

You have to practice listening, too. Learning to respect and collaborate with other workers on the job is essential. You never know who on a job site will come up with a great idea.

I think this is one of the best careers available today. There are opportunities everywhere. A good electrician never has to be out of work for long. You can almost always find work doing something. It is not all about construction. We are needed in numerous industries, from steel mills to film and television production.

An electrician can specialize, too. Doing maintenance for factories and other commercial ventures is a good way to go if you want stability. There are many more subsets. You could be a gaffer and coordinate the lighting on a movie set. You could retrofit electrical systems for video displays in historical buildings, or you could hook up alternative energy sources in homes and businesses who want to rely on green energy.

No matter what kind of work you choose to do, you have to be good at it and you have to enjoy it. Getting a job is easy. Holding a job will be your first challenge. You have to want it and like it. If you have a drive for it, you won't have any trouble moving along in your career."

I Am a Solar Electrician

"I didn't set out to become an electrician. I was studying conservation and ecology in college when the subject of solar power caught my interest. I had watched solar panels pop up on the roofs of homes in my Colorado neighborhood and it got me curious. I took a few classes in the field of solar energy and got into an internship with Solar Energy International. I learned a lot about renewable energy there and my passion for solar grew.

My first job offer came from a local solar company right after graduation. I jumped on it because they offered paid on-the-job training while I worked towards getting my Limited Renewable Technician license (LRT). It is the license that allows me to legally install solar electric systems.

I am proud to work every day to better the environment. My work involves everything from designing PV (photovoltaic, meaning light electricity) systems to doing full installations. About half my day is spent outside working on systems. The rest of the time is in the office working on the computer or sitting in on meetings to plan upcoming projects. I also do site surveys, which involves going to a potential customer's property and analyzing the location to see if it is a good match for a solar power system.

I love this work. I'm not cooped up inside, and there are opportunities for travel, which makes me happy. Mostly, I like that I can truly feel good about my work at the end of the day. I'm doing my part to better the environment and I am educating others on how they can help. If you want to do something meaningful with your life, you should consider it. It's an up and coming industry with loads of opportunity."

I Am a Maintenance Electrician

"My job is to test electrical equipment to make sure it's working properly and is safe. I started out working for a tech manufacturer, looking after their security systems. After I got married and had kids, I wanted more money and flexibility so I went into business for myself.

I worked in construction for years before becoming an electrician. I was a builder, but I was always particularly good with electrical equipment. I could take one look at the blueprints and pinpoint flaws in the proposed electrical system. There are always more efficient ways to wire and I have a knack for seeing them.

My first employer paid for my schooling to qualify as an electrician. It was hard going, especially the math part. I remember struggling with high school algebra and thinking it couldn't possibly have anything to do with my life. Now I use it every day. It's surprising how much math is involved.

The best thing about being an electrician, especially a self-employed electrician, is the variety. There is no typical workday. I am always out and about, not chained to a desk all day. I also like the challenge of troubleshooting when problems come up.

What I dislike most is the paperwork. There isn't much paperwork for most electricians, but running my own business is a whole new layer of responsibility. The business side of things doesn't come naturally to me and managing a team of people is challenging, but it's worth it to be in control of my time and income.

My best advice to anyone considering this field is to always take pride in your work. Finding a job is easy. Simply find your local electrical workers union and see what's available for beginners. Keeping the job and moving up requires the respect of your employers and your peers. Work hard and learn everything you can, and you will be rewarded with good pay and a stable, long-lasting career."

PERSONAL QUALIFICATIONS

SUCCESSFUL ELECTRICIANS HAVE several characteristics and traits in common. For example, they have technical aptitude and they are keenly interested in math and science, particularly physics.

All electricians should be in good health. The work can be strenuous and it often demands stamina to remain in awkward positions for extended periods of time. Constant standing and kneeling can be tiring! Although physical strength is needed to carry the tools and supplies of the trade, agility and dexterity are equally important. Be prepared to work in cramped spaces or on scaffolding.

Color vision is essential. Color-blind people cannot become electricians because all wiring is color-coded to avoid mistakes and injuries.

Do you enjoy working with your hands? The work of an

electrician is very much hands-on and good hand-to-eye coordination is a must.

Being detail oriented is a big plus. Working with electricity has inherent risks. To reduce the possibility of accident or injury, an electrician must pay attention to details at all times. A careful approach is crucial for insuring that electrical components are installed in compliance with safety codes.

The ability to concentrate for long periods of time is extremely important. You must come to work rested, and stay focused and alert throughout the day. The consequences of poor quality work can be enormous. Mistakes often result from being tired.

Electricians are troubleshooters who find, diagnose, and repair problems. When a motor or outlet stops working, for example, it becomes necessary to perform the right tests to determine the cause of the failure. Critical thinking skills are needed to use the results to determine the best course of action to fix the problem.

Interpersonal skills are helpful. Most electricians work with people of all kinds on a regular basis. They should be friendly and able to answer questions from customers and other contractors.

This field is always advancing. Every day brings new techniques, applications for power usage, shifts in power sources, and changes to safety codes. The most successful electricians are those who are eager and willing to learn something new. Taking classes and improving your knowledge will pay off in more job opportunities and higher wages.

ATTRACTIVE FEATURES

THE CHOICE TO BECOME AN ELECTRICIAN has good merit. Most people who make the decision are happy with their career and stay in the field all of their working lives. There are many reasons people move into this field, but the five best long-term benefits are listed below.

Solid salary

The number one benefit of becoming an electrician is the potential for high earnings. An electrician's actual salary depends on the type of job, number of years on the job, and specialty, if any. Across the board, qualified electricians earn a median annual salary of $50,000 plus attractive benefits. Beginners start out with a good living wage, and the knowledge that they could eventually have the necessary skills and experience to earn more than $80,000. Along the way, it is possible to boost earnings by working overtime, which typically pays time-and-a-half.

There are also opportunities for independent-minded people who want the option to earn more than an employer might want to pay. Opening up one's own shop can lead to an income that is potentially much higher than what you can earn working for a company. However, self-employed workers have to cover their own benefits, including health insurance.

Easy entry

The income looks even better when considering how easy it is to enter the field. A college education is not necessary, though a degree in electronics would certainly be a valuable asset. Most electricians start out as apprentices, and learn on the job while earning a paycheck. The minimum age requirement for most apprenticeship programs is 18. Some electricians even manage to start their careers straight out of high school. Whether that is possible depends on what training resources are available at a particular high school or in the area.

Job security

There is a growing shortage of qualified electricians. The demand is high for those with the necessary skills to do installation, maintenance, and repair work on a variety of electrical projects. This is good news for anyone looking to get their first job. It is even better for working electricians who enjoy the secure feeling of being in a stable career. Becoming an electrician is signing up for a lifelong career since electrical systems will always need to be installed and maintained by professionals.

Flexible career choices

Electricians can take advantage of a diverse and changing work environment with various kinds of projects. The typical electrician can easily move from outdoor projects such as laying cables and installing street lighting to indoor work, from residential construction to wiring huge commercial buildings. You can choose to work only in the commercial, housing, or industrial sector, and then change for any reason. An electrician can work as an employee of a larger business, or start a small business and become self-employed. There are also specialty options to consider. Among the many possibilities are systems maintenance, mining, water supply, renewable energy, and vehicle manufacturing.

More than a job

Being an electrician is a career choice, not just a job. The work requires special training and the ability to meet the demands of a challenging field that is experiencing rapid technological change. Electricians are expected to keep up with industry changes. As they do so, many learn of new opportunities in emerging specialties that may require more education and training. As in most careers, there is also the opportunity for upward mobility. Management positions exist for those with proven trade skills as well as experience dealing with professionals at all levels.

UNATTRACTIVE FEATURES

THE WORK OF AN ELECTRICIAN CAN BE PHYSICALLY demanding. Generally, electricians have to climb ladders, be on their feet for much of the day, and lift tools and supplies – some of which can be quite heavy. They may have to bend, squat, or kneel to make wiring connections, often in awkward locations. They may be in a cramped crawl space or high up on scaffolding.

Electricians work in a variety of conditions, which may expose them to the elements. Those making emergency repairs to power systems often work outdoors where they are subjected to all kinds of weather conditions. Those working in construction or industrial plants are usually protected from the weather, but their work sites can be very noisy and dirty.

An electrician's work is potentially hazardous. Although few accidents are fatal, electrical shocks, burns, falls from scaffolding, and cuts from sharp tools are possible. Electricians also wear specially designed protective clothing such as gloves with insulating rubber liners, safety glasses to protect against arc flash exposure, flash-resistant coveralls, and work boots and hard hats that are specially rated to provide protection from shock and mechanical impact.

There are hazards associated with any trade, but working with electricity poses particular risks in the workplace. Electricians are trained to adhere to strict safety guidelines to reduce risks. This training keeps electricians in the minority when it comes to electrocutions. Only a small number of construction workers who are electrocuted are electricians. Most are other workers who have not benefitted from the same training that teaches how to safely deal with electrical systems and components.

Most electricians work normal hours, but there are exceptions. Some have to be on call in case problems arise. Those working in construction may be needed to work overtime to complete a project. Maintenance jobs often have to be performed during evenings or weekends when commercial facilities are closed. Maintenance has to be done according to a regular schedule that may overlap holidays. Shift work is common in large facilities that have to operate around the clock. To make sure someone is always present, there are three different shifts: day, swing, and graveyard (overnight). In most cases, workers rotate through the shifts so no one is stuck working nights all the time, but it can be difficult adjusting to changing work schedules.

EDUCATION AND TRAINING

ELECTRICIANS CAN LEARN THEIR trade in several ways, none of which requires a college degree. Most learn the ropes through an apprenticeship, by attending a technical school, or working as an electrician's assistant.

Some do go to college though, typically to obtain a degree in electronics or electrical engineering. This kind of education, while not required, creates a greater number of unique career opportunities that can make you more money. It can also accelerate your career as an electrician. For example, most states require anyone taking the master electrician exam to have completed seven to nine years of experience as a journeyman. However, a Bachelor of Science degree qualifies you to take that same master exam as soon as you become a journeyman. Even considering the time spent in school, that shortcuts the time frame by as much as five years. This paves the way to rapid advances to high-paying planning and supervising positions. A degree makes you eligible for other higher-level careers in the electrical field, such as electrical technician and electrical engineer. These professionals are highly sought after in the industrial sectors where they work in power plants, car manufacturing plants, and textile producers.

On-the-Job-Training

Regardless of how you learn the trade, training and experience are essential. Most electricians go through an apprenticeship program, but those who do not, usually begin as assistants to experienced electrical contractors. They learn many of the same things as apprentices, but may not receive training in as many areas because such on-the-job training is often specific to a single type of work. Many trainees are simultaneously enrolled in vocational or correspondence schools.

Apprenticeships

Several groups, including unions and contractor associations, sponsor apprenticeship programs. Programs are conducted under the general supervision of a master electrician and usually under the direct supervision of a journeyman electrician. Programs may vary, but the basic qualifications to become an apprentice usually include:

A high school diploma or equivalent

Minimum age of 18

One year of algebra

Qualifying score on an aptitude test

Pass substance abuse screening

Some apprentice electricians start out by attending a technical school. Many technical schools offer programs related to circuitry, safety practices, and basic electrical information. Graduates usually receive credit toward their apprenticeship.

Apprenticeship programs last from three to five years. For each year of the program, apprentices must complete at least 144 hours of technical training in the classroom and 2,000 hours of paid on-the-job training. In the classroom, apprentices learn electrical theory, blueprint reading, mathematics, electrical code requirements, and safety and first-aid practices. They may also receive specialized training in related areas such as soldering, communications, fire alarm systems, and elevators.

On the job, apprentices get practical training by working as assistants to experienced workers and then progress to performing jobs themselves. They may start with simple tasks like drilling, placing conduit, and setting supports. As they gain skills and experience, they are given more responsibility. Before finishing an apprenticeship, they will be able to measure, fabricate, and install conduit, as well as install, connect, and test wiring, outlets, and switches. They will also know how to set up and draw diagrams for entire electrical systems.

For those interested in becoming maintenance electricians, a background in electronics is increasingly important because of the growing use of complex electronic controls on manufacturing equipment. These individuals must obtain this knowledge through higher education in a vocational school or college.

Advancement

After completing an apprenticeship program, electricians are considered journeymen and may perform duties on their own, subject to any local licensing requirements. Journeymen continue to get more experience and begin to learn more specialized areas like low voltage installation, telecommunications, and audio-visual systems.

Experienced journeymen electricians can become supervisors, managers, or superintendents. Those with sufficient capital and management skills may start their own contracting business, although this may require an electrical contractor's license. Others might become building inspectors who specialize in electrical systems.

The highest level for an electrician is the master. Journeyman electricians can work unsupervised provided that they work according to a master's direction. Most states do not offer journeyman permits, and therefore they must work under permits issued to a master.

Licensing

Most states have their own licensing exams that test knowledge of local regulations as well as information contained in the National Electrical Code, the national register of electrical regulations. Exact requirements vary by state, but all states recognize three basic skill categories: master, journeyman, and apprentice trainee.

Applicants need to take a test that assesses how well they know the National Electrical Code, local building codes, and theory. The tests have questions related to the National Electrical Code, state electrical codes, and local electrical codes. For more information, contact your local or state electrical licensing board.

Electricians periodically take continuing education courses offered by their employer or union to keep current about any changes or additions in the National Electrical Code, new materials, or new methods of installation. These courses sometimes come directly from manufacturers who provide training in procedures for specific products.

EARNINGS

THE MEDIAN ANNUAL EARNINGS FOR electricians are about $50,000. That is a respectable income for a career that does not require a college education. Even the lowest paid electricians do okay at around $30,000, while the top 10 percent earn more than $85,000.

It is common for electricians, like all tradesmen, to be paid an hourly wage. Nationwide, that works out to about $22.50 per hour. Almost all electricians work full time though, which sometimes includes evenings and weekends. It is common for those working on scheduled maintenance to put in overtime. At time-and-a-half for overtime, that is a big plus rather than a problem. Overtime is also common on construction work sites, where meeting deadlines is critical.

The lowest paid electricians are usually still apprentices. The starting pay for apprentices is typically between 30 and 50 percent of what fully-trained electricians make. As apprentices gain skills, they receive periodic pay increases to reflect their level of experience.

Earnings vary by specialty, industry, location, and union versus nonunion. Electricians working in nonresidential building construction earn close to the national average, around $22 an hour. Those who deal with large installations and maintenance, such as electric power generation, or transmission and distribution are considered specialists. They are rewarded for their expertise with earnings of nearly $30 per hour.

The best-paying industries include natural gas distribution and motion pictures, where electricians are known as gaffers. The four cities offering the highest pay include San Francisco, New York City, Fairbanks, and Trenton.

Union workers generally earn more than nonunion workers. That may account for the fact that one third of all electricians in the US belong to a union – that is a higher percentage of unionization than almost any other occupation. Unions use their collective bargaining power to negotiate contracts for jobs or with specific employers. Pay rates are an integral part of the contract.

Most electricians belong to the International Brotherhood of Electrical Workers (IBEW), though there are other unions for electricians. According to the IBEW, union jobs overall pay between 25 and 30 percent more than nonunion jobs for the same work. As with nonunion work, wages may vary depending on the industry. For electricians working in the construction industry, the difference is significant. In that arena, the IBEW reports 50 percent higher wages. Electricians in the telecommunications or utilities industries do not do nearly as well, but they still could earn as much as 12 percent more than their nonunion counterparts.

OPPORTUNITIES

THE FUTURE IS BRIGHT FOR SOMEONE who chooses this career path. Employment of electricians is projected to grow 20 percent over the coming decade. That is faster than the average for most other occupations. New electricians will find numerous openings arising each year due to many experienced electricians reaching retirement age, changing careers, or leaving the occupation for other reasons. Currently, there are not enough potential workers to replace them because many prefer work that is less strenuous and has more comfortable working conditions. That means there will be less competition for those who are willing to fill the void.

Job growth can also be attributed to the growing population that needs electricians to install and maintain electrical devices and wiring in homes, factories, offices, and other structures.

Well-trained workers will have the most favorable

opportunities – especially those who are handy with electronics. Innovations in technology are creating greater demand for electricians as buildings need to have electrical systems for computers and advanced communications, including cable TV and the Internet. More factories are using automated and robotic equipment. Existing buildings need to be remodeled and retrofitted with electronic systems.

Alternative power generation, using solar and wind, is an emerging field that promises a great future for electricians who are ready to adapt to green power generation. Electricians who specialize in environmentally friendly technology will be needed to link alternative power sources to privately owned buildings as well as public power grids for many years to come.

Employment of electricians tends to fluctuate with the overall economy. This is seen mostly in the construction industry, which is susceptible to the up and down rate of new building. The fact that most jobs are project based makes it easy for employers in the construction industry to cut back during downturns. There might also be fewer openings in apprenticeship programs during those times. Upswings in the economy can make it equally difficult to find openings. Turnover for electricians in construction is lower than other construction occupations because the income is generally higher. All of this may seem like construction is an unstable industry to get into. However, there are so many electricians from the baby boom generation who are reaching retirement age, a net increase in job opportunities is expected.

One of the most stable areas for electricians is maintenance. Unlike construction, which can start and stop at any time, maintenance of existing systems must continue even when there is little new construction.

Therefore, maintenance electricians are less subject to layoffs and slow seasonal hours than those working in construction and manufacturing. On the other hand, some companies will try to cut costs by using contracted electricians rather than by employing in-house electricians. This may be bad news for salaried employees, but the trend will increase opportunities for electricians who work for contracting companies. One way or another, electrical work needs to be done.

GETTING STARTED

ALL ELECTRICIANS START OUT AS apprentices. They may not be called that if they are not hired by a union employer, but it simply means they go through some sort of formal or informal training on the job. There are basically two ways to get an apprenticeship: through the local union or directly from an employer.

Getting into a union apprenticeship program is a fairly straightforward process, though not always an easy one. Union apprenticeship programs are often administered by local management committees made up of union members and local employers. Any high school graduate who is at least 18 years old and in good physical condition can apply for apprentice jobs. You will be asked to fill out an application and be interviewed like any other job. There is a test, but it is an aptitude test that will not necessarily rule you out as a qualified applicant if you do not score high. If you have not attended a technical high school, it is a good idea to take an online course that will at least familiarize you with the basics.

If you have a friend or relative in the union, be sure and

use that name as a reference. There was a time when having a relative in the union was a requirement for entry. That is no longer the case, but old traditions still linger and you can use them to your advantage.

Applying directly to potential employers takes a little more legwork and commitment, but there is nothing complicated about it. Start by browsing job boards on the Internet. Look on Craigslist, general job boards, and on specialty sites that cater to electricians and general contractors. It is common for electricians, even those just starting out, to find their first jobs through word-of-mouth. Let everyone know you are looking for this kind of work. Do some basic research to find local building contractors and give them a call. If they are not hiring any trainees, ask if they know of anyone who is. It is a small world. Companies usually know what other companies are doing. This can be a very effective way of getting to the right company quickly. Be sure and use the referrer's name. Having a name to use breaks the ice and gets attention.

Many employers send job opening announcements to vocational-technical schools. Taking electrician courses in a vo-tech school is a good way to get you started. Make a point to get to know your teachers. They are usually experienced electricians themselves and very often can be a great resource for introducing you to potential employers. Take advantage of any job placement services offered at your school.

ASSOCIATIONS

■ **National Electrical Contractors Association (NECA)**
http://www.necanet.org

■ **Associated Builders and Contractors**
http://www.abc.org

■ **Independent Electrical Contractors**
http://www.ieci.org

■ **National Association of Home Builders**
http://www.nahb.org

■ **International Brotherhood of Electrical Workers (IBEW)**
http://www.ibew.org

■ **National Association of Home Builders, Home Builders Institute**
http://www.hbi.org

■ **National Center for Construction Education and Research**
http://www.nccer.org

PERIODICALS

■ **Electrical Contractor Magazine**
http://www.ecmag.com

■ **Electrical Business Magazine**
http://www.ebmag.com

WEBSITES

■ National Joint Apprenticeship Training Committee
www.electricaltrainingalliance.org

■ The National Center for Construction Education and Research
http://www.nccer.org/about

■ Electrical Agent
http://www.electricalagent.com

■ Electrical Jobs Today
http: www.electricaljobstoday.com